GUINNESS WORLD RECORDS

Fantastic Flight Records

Compiled by
Laurie Calkhoven and Ryan Herndon

For Guinness World Records:
Laura Barrett, Stuart Claxton, Craig Glenday, David Hawksett,
Hein Le Roux and Ben Way

SCHOLASTIC INC.
New York Toronto London Auckland Sydney
Mexico City New Delhi Hong Kong Buenos Aires

The publisher would like to thank the following for their kind permission to use their photographs in this book:

Cover: © NASA; 1: © Drew Gardner/REX USA; 2: © Bettmann/CORBIS; 3: © Hulton-Deutsch Collection/CORBIS; 4: © Bettmann/ CORBIS; 5: Hulton-Deutsch Collection/CORBIS; 7 © Bettmann/ CORBIS; 8 (top): © Georges de Keerle/Sygma/ CORBIS; 8 (bottom): © Jim Sugar/CORBIS; 9: © Reuters; 10: APTV/AP Wide World Photos; 11 (top): © Douglas Slone/CORBIS; 11 (bottom): © Natasha Lane/AP Wide World Photos; 12: © Corbis; 13, 14: Getty; 15: Hulton-Deutsch Collection/CORBIS; 16: Bettmann/CORBIS; 17: © Bill Ross/CORBIS; 18: © Bruce Burkhardt/CORBIS; 19: © Airsport Photo Library; 20: © Kevin Fleming/CORBIS; 21: (© E. James/NASA; 22: © Alain Nogues/Sygma/CORBIS; 23: © Reuters/CORBIS; 24, 25: © NASA; 26: © NASA/Roger Ressmeyer/CORBIS; 27: US Airforce Photo; 28: © Ronald Martinez/AP Wide World Photos; 29 © Stock Trek/Photodisc/PICTUREQUEST; 30: © Bettmann/CORBIS

ISBN 0-439-71566-0

Cover design by Louise Bova
Interior design by Two Red Shoes Design Inc.
Photo Research by Els Rijper
Records from the Archives of Guinness World Records

12 11 10 9 8 7 10 11 12 13 14/0

Printed in the U.S.A. 40

First printing, February 2005

Visit Scholastic.com for information about our books and authors online!
www.guinnessworldrecords.com

People have always
wanted to fly like birds.

This *Guinness World Records*™ book takes you
from the earliest days of flight to the latest
rocket blasting off into space.

Soar into the sky with these
record-holders!

Airplanes are *heavier* than air. They need wings to lift them into the sky and engines to keep them up there. They also need brave people to be their pilots.

RECORD 1

First Power-Driven Flight

Brothers Orville and Wilbur Wright believed that flying an airplane was like riding a bicycle. A plane needed a rider, or **pilot**, to steer where it would fly.

The Wright brothers built special kites and tested different wings in a homemade wind tunnel. They built an engine and a propeller, and found the perfect place to test their plane — at Kill Devil Hills near the town of Kitty Hawk in North Carolina. On December 17, 1903, they were ready for the big test.

The brothers took turns in the pilot's seat. Orville flew first for 12 seconds. His **altitude** — how high something is above the ground — was between just 8 and 12 feet.

The brothers made history (see photo) by following their dreams. Today, *Flyer I* is at the Smithsonian in Washington, D.C.

Flying Solo

A solo flight is when one person flies the plane alone.

On July 15, 1933, pilot and daredevil Wiley H. Post (see photo) took off in his plane, *Winnie Mae*, from New York City. Seven days, 18 hours, and 49 minutes later, he landed back in New York City. Wiley was the first person to fly solo around the world!

RECORD 2

Largest Wingspan

Wingspan is the distance between the wingtips. The largest wingspan is 319 feet 11 inches — longer than a football field — and it is on a flying boat!

Flying boats or **seaplanes** can take off and land on water. During World War II, the U.S. military used seaplanes to send soldiers safely across the Atlantic Ocean.

A rich man named Howard Hughes (photo, top right) built the biggest flying boat ever. The plane was nicknamed the *Spruce Goose* (see photo). It had eight engines and could hold up to 700 soldiers!

The war ended before the giant seaplane was ready. People said it was too big to fly. Hughes proved them wrong. On November 2,

1947, he piloted the *Spruce Goose* and flew it for 60 seconds. Then he put the plane back in the hanger and it never flew again.

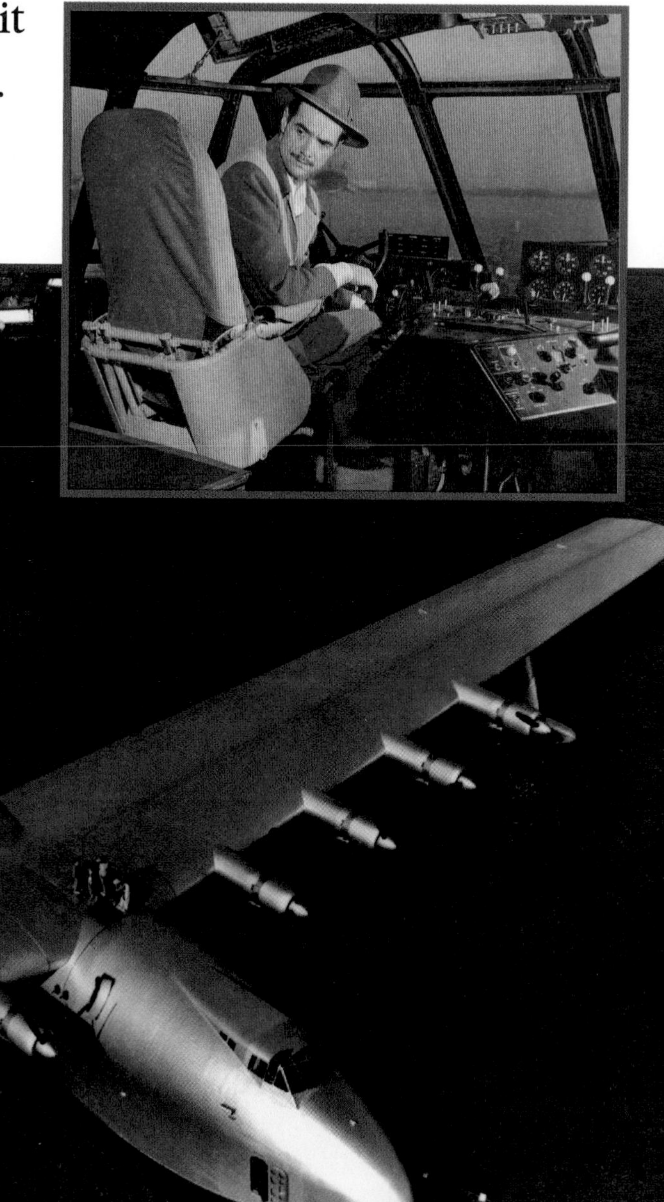

Sky Dance

At air shows, you see pilots make their planes dance in the sky. These loops, rolls, and spins are known as **aerobatics**.

The first skill these pilots learn is how to fly upside down — without getting sick! Joann Osterud from Canada flew 4 hours, 38 minutes, and 10 seconds to set the Guinness World Record for the longest upside-down flight.

Every two years, teams from around the world compete in the World Aerobatic Championship. Judges score the pilots on their skills. The U.S. men's team has won the championship six times. The Russian women's team has won four times.

UP, UP, AND AWAY!

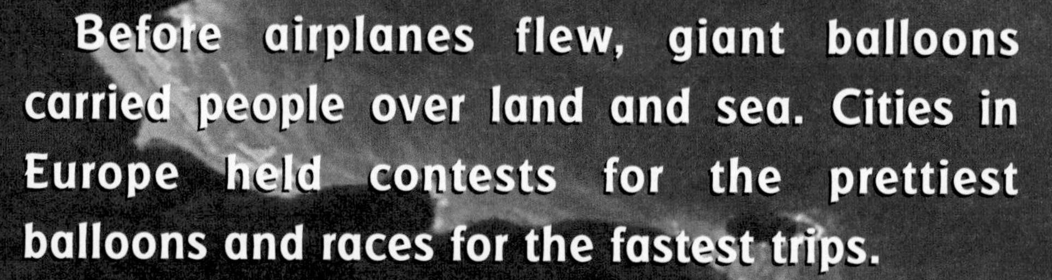

Before airplanes flew, giant balloons carried people over land and sea. Cities in Europe held contests for the prettiest balloons and races for the fastest trips.

Pilots can only move balloons up and down. So how do they get from place to place? They ride the wind. The wind blows fast or slow at different places in the sky. The higher up you go, the faster your balloon flies.

RECORD 3

Longest & Fastest Solo Balloon Flight Around the World

American Steve Fossett (see photo) knows which way the wind blows. During the summer of 2002, Steve flew his balloon around the world in 14 days, 19 hours, and 50 minutes — the fastest trip and the longest time anyone has spent in a balloon alone.

This was not Steve's first attempt. In 1997, he crashed in Russia. In 1998, a storm caused his balloon to fall into the Coral Sea. In 2001, more bad weather forced him to land in the middle of a cattle farm in Brazil.

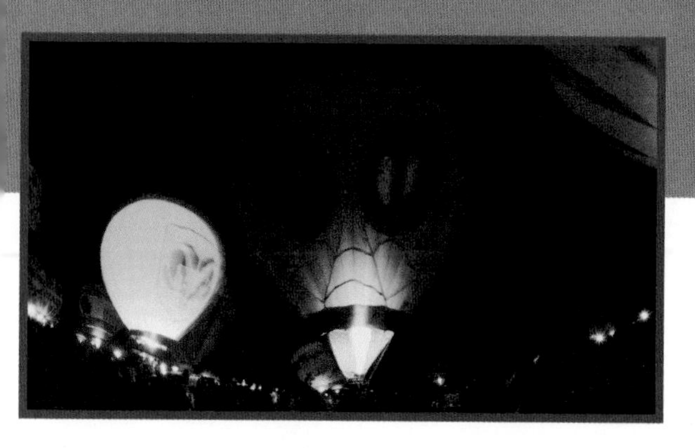

Every year, a million people go to the Balloon Fiesta in Albuquerque, New Mexico.

Many people work together to get their hot-air balloons into the sky at the same time. During one hour on October 7, 2000, the balloons — all 329 of them — made their **ascent**, or liftoff (see photos).

Balloons have many uses — for fun, travel, and defense. Soldiers rode in the balloon's basket, or **gondola**, and took photos of enemy camps during the Civil War. In World War II, the British tied giant balloons around London, England, to stop low-flying enemy planes from bombing the city (see photos).

People also traveled in a giant balloon called an **airship** or a **dirigible**.

An airship is *lighter* than air. It is a big balloon filled with a special gas — either hydrogen or helium. Helium is the same gas that makes birthday balloons rise up into the sky.

There are different kinds of airships. A **blimp** has a bendable outer "skin." **Zeppelins** have hard, stiff skins. Today, blimps fly over baseball and football games.

The world's largest working airship today is the WDL 1B. This airship is 197 feet long — about the same length as an entire city block! The photo above shows the airship's **frame** or skeleton. Once the outer covering is sealed, the airship is ready to go.

RECORD 4

Largest Airship Ever

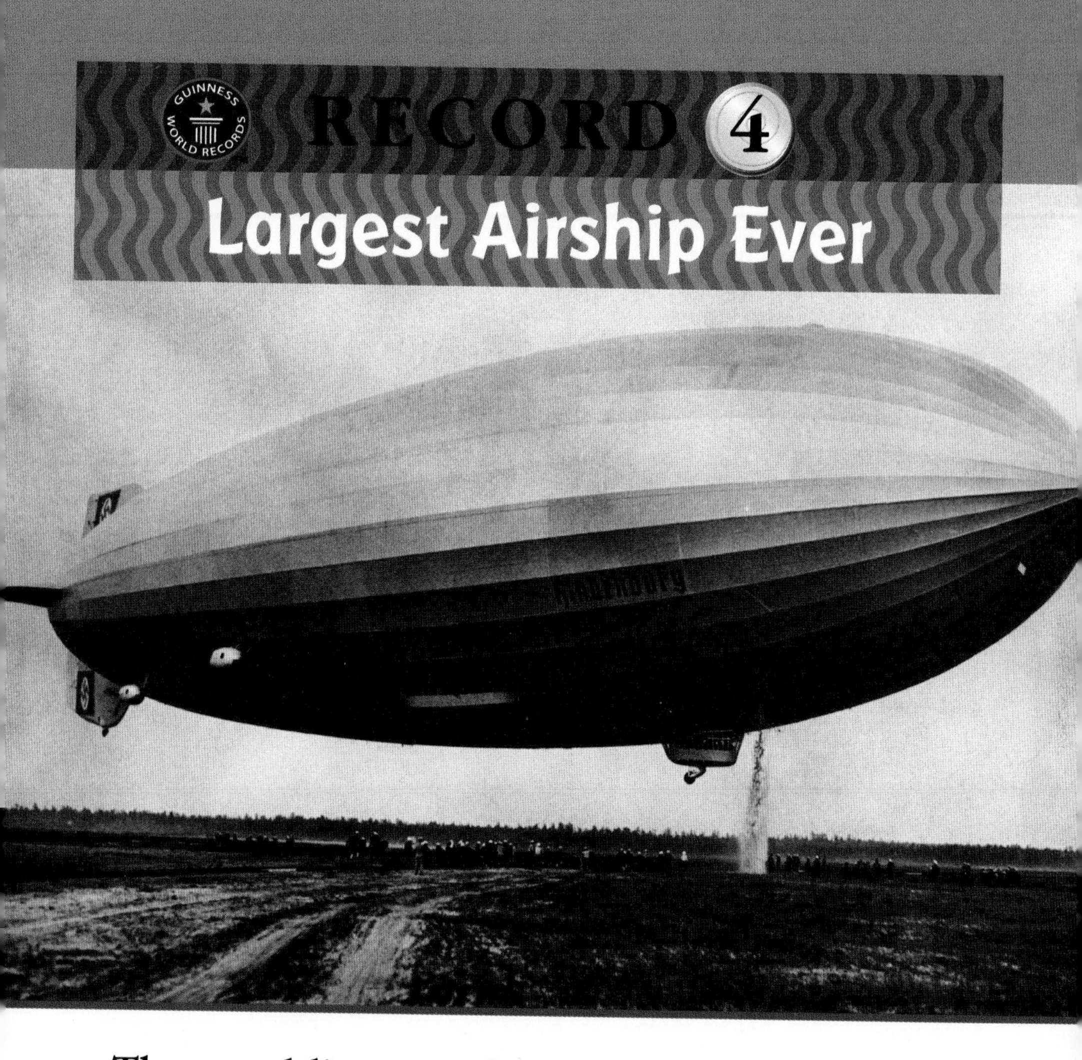

The world's most famous airships were the *Hindenburg* (see photo) and its sister ship *Graf Zeppelin II*. Each one was 803 feet 10 inches — that's as long as ten blue whales!

In 1936, people did not know hydrogen gas was dangerous. Now airships use helium gas. The *Hindenburg* used hydrogen gas to make 10 flights across the Atlantic Ocean from

Germany to New Jersey. In 1937, the gas caught on fire and the ship exploded.

People feared airships were not safe. The *Graf Zeppelin II* flew in 1938, but never across the ocean. It was later broken up and used to build fighter planes.

Depending on the weather and wind, the airship's flight across the Atlantic took about two days. Today, a jet plane can make the same trip in less than six hours!

FLY LIKE AN EAGLE

Do you want to feel the wind beneath your wings? If so, fly gliders or a hang glider!

The first kind of plane the Wright brothers built was a glider. Gliders are like paper airplanes. They do not have engines. Their wings are long and narrow to help them ride the wind.

You **launch** a paper airplane (see photo) by throwing it into the air. Gliders use a **winch**, or giant slingshot, to throw them into the air. Airplanes also tow the glider into the sky. A long rope ties them together. When the glider is high enough, the pilot signals and the rope is unhooked. Then the glider soars through the sky!

RECORD 5
Greatest Altitude in a Glider

Robert Harris (see photo) took a solo flight in his glider on February 17, 1986. He set a Guinness World Record by soaring to an altitude of 49,009 feet — more than nine miles above the earth!

Sabrina Jackintell flew her glider on February 14, 1979. Her record altitude was 41,460 feet. No woman has soared higher than Sabrina in more than 25 years!

A hang glider is like a giant kite. A harness straps the pilot to a trapeze-like frame. Pilots use their body weight to steer.

Hang glider pilots need enough air underneath the kite to lift off. They take off by running down hills and leaping into the air. Small planes also tow the hang glider into the sky. The pilot unhooks the rope holding him to the plane and then flies away (see photo)!

ROTORS AND ROCKETS

The Wright brothers taught us that airplanes can fly. Helicopters were the next big invention. The first piloted helicopter flew in 1907!

Unlike airplanes, helicopters can take off and land *vertically*. They fly straight up and down, just like balloons.

RECORD 6

Biggest Helicopter

The Russian Mil Mi-26 (see photo) is 131 feet long, and as tall as four men standing on one another's shoulders. The tail rotor has a diameter of 25 feet. This helicopter is big enough to hold 100 troops inside, and weighs up to 123,500 pounds. The Mil's top speed is 183 miles per hour.

Jet engines are more powerful than rotors or propellers. The engine pulls air into the front. Air and fuel mix and explode. This explosion burns fuel. Leftover hot gases pushed out of the back move the jet forward — fast!

A flight faster than the speed of sound is **supersonic**. When an aircraft breaks the sound barrier at a low altitude (see photo), shock waves cause a **sonic boom**. People on the ground hear a huge "boom!"

RECORD 7
Fastest Jet

Mach numbers measure supersonic speed. At Mach 1, an airplane is flying at the speed of sound. The U.S. Air Force's Lockheed SR-71 *Blackbird* (see photo) flies at speeds over 2,200 miles per hour — Mach 3, more than three times the speed of sound!

NASA now uses this jet for high-altitude experiments. It can fly at altitudes of 100,000 feet, but the pilots wear the same space suits as astronauts do to be able to breathe!

RECORD 8

Highest-Flying Propeller-Driven Aircraft

From a distance, *Helios* looks a bit like a flock of birds because of its 14 propellers (see photo). This NASA aircraft reached an altitude of 96,500 feet, higher than 18 miles above the earth!

Because its motors get energy from the sun, *Helios* does not need fuel and may be able to stay in the air for months. *Helios* is an **unmanned** aircraft, which means it does not need a pilot onboard. It can also fly at an altitude three times higher than Mount Everest!

BLAST OFF

Jets can only get you so high. So what if you want to travel into space? Jets don't work in space because there is no oxygen. Rockets, however, take their own oxygen in *fuel tanks*, allowing them to go farther.

Rocket-powered spacecraft use more than one set of engines during a mission. Each set is called a *stage*. The space shuttle (pictured) uses special engines called *boosters* that sit on the outside of the fuel tank. These boost, or push, the rocket forward.

RECORD 9

Smallest Rocket

Pegasus is the smallest rocket that can send cargo into space. It is a three-stage booster rocket and only 50 feet 10 inches long.

Most rockets blast off from a launch pad. *Pegasus* hitches a ride on a jumbo jet named *Stargazer*. This jet carries the rocket (see photo) to an altitude of 40,000 feet — as high as most planes can fly. *Pegasus* detaches itself from *Stargazer*. Then it fires its first-stage engine and heads for outer space!

NASA

The National Aeronautics and Space Administration (NASA) is the U.S. government agency that explores space.

In 1981, NASA launched the first space shuttle, *Columbia*. This aircraft is used to carry people and equipment into outer space – and back!

RECORD 10

Longest Extraterrestrial Balloon Flight

Can you believe that we use balloons to explore other planets?

Twin Soviet spacecraft *Vega 1* and *Vega 2* launched in December 1984. Six months later, they reached Venus (see photo). They dropped packages into the planet's atmosphere. Each package split into a **lander** and a balloon to send back important information about Venus. The balloons only lasted for a few hours before being destroyed by the harsh atmosphere. After visiting Venus, the probes went on to make a flyby past Halley's Comet in March 1986.

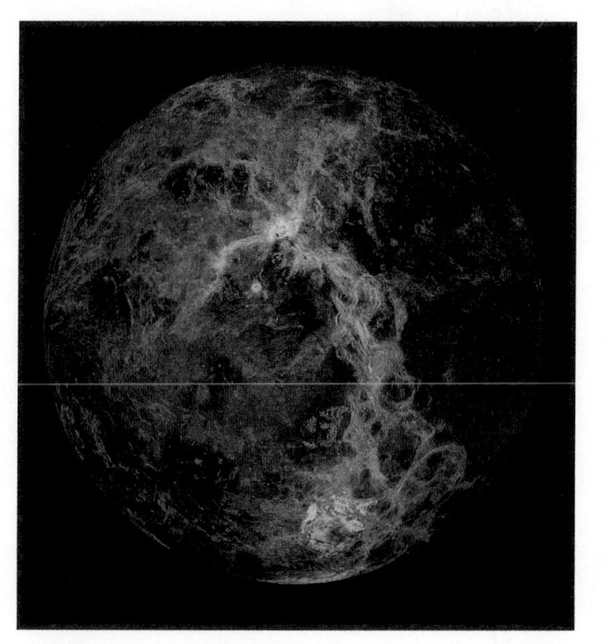

What's next in the fantastic history of flight? People will always want to travel farther and fly higher. Maybe one day you'll take a vacation in outer space . . . go gliding on Venus . . . or zip through the air on your own set of rocket-propelled wings. Maybe you will even break the next Guinness World Record in the sky!